ISBN 978-1-334-48162-8
PIBN 10715679

English
Français
Deutsche
Italiano
Español
Português

www.forgottenbooks.com

Mythology Photography **Fiction**
Fishing Christianity **Art** Cooking
Essays Buddhism Freemasonry
Medicine **Biology** Music **Ancient**
Egypt Evolution Carpentry Physics
Dance Geology **Mathematics** Fitness
Shakespeare **Folklore** Yoga Marketing
Confidence Immortality Biographies
Poetry **Psychology** Witchcraft
Electronics Chemistry History **Law**
Accounting **Philosophy** Anthropology
Alchemy Drama Quantum Mechanics
Atheism Sexual Health **Ancient History**
Entrepreneurship Languages Sport
Paleontology Needlework Islam
Metaphysics Investment Archaeology
Parenting Statistics Criminology
Motivational

U.S. DEPARTMENT OF COMMERCE
NATIONAL OCEANIC AND ATMOSPHERIC ADMINISTRATION
NATIONAL MARINE FISHERIES SERVICE

NOTE

Until October 2, 1970, the National Marine Fisheries Service, Department of Commerce, was the Bureau of Commercial Fisheries, Department of the Interior.

UNITED STATES DEPARTMENT OF COMMERCE
Maurice H. Stans, Secretary

NATIONAL OCEANIC AND ATMOSPHERIC ADMINISTRATION
Dr. Robert M. White, Administrator

NATIONAL MARINE FISHERIES SERVICE
Philip M. Roedel, Director

Oil Pollution on Wake Island from the Tanker R. C. Stoner

By

REGINALD M. GOODING

Special Scientific Report--Fisheries No. 636

Seattle, Washington
May 1971

For sale by the Superintendent of Documents, U.S. Government Printing Office
Washington, D.C. 20402 - Price 25 cents   Stock Number 0320-0008

FIGURES

iii

Oil Pollution on Wake Island from the Tanker R. C. Stoner

By

REGINALD M. GOODING, Fishery Biologist

National Marine Fisheries Service
Hawaii Area Fishery Research Center
Honolulu, Hawaii 96812

ABSTRACT

On September 6, 1967, the tanker R. C. Stoner foundered on the reef off the harbor entrance at Wake Island. During the following 10 days the vessel's cargo of over 22,000 kliters (6 million gal) of high octane aviation gasoline, aviation jet fuel, aviation turbine fuel, diesel oil, and bunker C black oil was spilled along the southern coast of the island.

A shore and underwater survey of the contaminated coastline showed that an estimated 2,500 kg of inshore reef fishes were killed and stranded on the shore. Numerous other fish and invertebrates were probably killed. Evidence is cited which indicates that most of the kill occurred on the shallow reef flat and the author speculates on the lethal effect of the various fuels.

INTRODUCTION

On September 6, 1967, the SS R. C. Stoner, an 18,000-ton tanker, went aground about 200 m southwest of the harbor entrance at Wake Island (Figures 1 and 2). She was in the process of mooring to two buoys located outside the harbor when the strong southwesterly wind drove her onto the reef.

When she foundered, the R. C. Stoner was loaded to capacity with over 22,000 kliters (6 million gal) of petroleum products. The cargo consisted of (1) 13,300 kliters (3,507,000 gal) of J-P4 military aviation jet fuel, a mixture of kerosene and gasoline which is light yellow and mixes readily with water; (2) 6,700 kliters (1,785,000 gal) of A-1 commercial aviation turbine fuel, a kerosene fuel which is light brown, mixes readily with water, and is a relatively safe fuel to handle; (3) 1,600 kliters (420,000 gal) of 115/145 aviation gasoline, a highly combustible fuel for high performance reciprocating engine aircraft, which is light purple and contains tetraethyl lead; (4) 640 kliters (168,000 gal) of diesel oil, which is

light brown; and (5) 525 kliters (138,600 gal) of bunker C black oil, the vessel's engine fuel and least volatile of the petroleum products on board.

Ollie Custer and I, of the Bureau of Commercial Fisheries Biological Laboratory (now National Marine Fisheries Service Hawaii Area Fishery Research Center), Honolulu, arrived at Wake Island on September 13 to survey and assess the damage to marine fish, invertebrates, and birds caused by the petroleum spillage.

Wake Island, lat 19°18' N, long 166°36' E, lies about 300 miles north northwest of the northernmost of the Marshall Islands and is administered by the Federal Aviation Agency (FAA). It has a loran station operated by the Coast Guard and a Pacific Missile Range installation maintained by the Air Force. Wake consists of three islets forming an atoll enclosing a shallow lagoon. The total land area is about 6.5 km^2 with a maximum elevation of about 7 m. The atoll is about 7.2 km long northwest to southeast and about 3.2 km wide. The lagoon has an area of about 9.1 km^2 and a

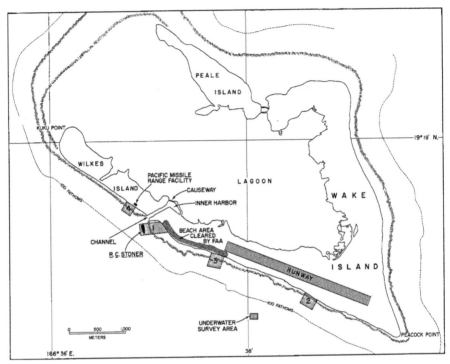

Figure 1.--Wake Island.

maximum depth of 3-4 m.

Immediately after the grounding, a large quantity of fuel spilled from the damaged tanks. It was believed to consist primarily. of aviation gas, J-P4 jet fuel, and possibly A-1 turbine fuel. However, the following day bunker C fuel was also escaping in considerable quantity (Figure 3). Gasoline vapor could be smelled until September 9, but not subsequently, indicating that the vessel was clear of gasoline within the first few days.

Unsuccessful attempts to pull the R. C. Stoner from the reef with her cargo intact made it apparent that it would be necessary to lighten the vessel before salvage operations could proceed. However, continuing southwesterly winds and rough seas caused by the recent close passage of typhoon Opal and tropical storm Rita delayed salvage. Southwesterly winds of 18 to 25 knots continued until the morning of September 16.

On September 8, the stern section of the R. C. Stoner broke off and plans to float the vessel free were abandoned. During the first 3 days after the grounding, the prevailing southwesterly wind drove the escaping fuel into the small boat harbor and along the coast for about 4 km towards Peacock Point (Figure 1). It was estimated that as much as 2,285 kliters (600,000 gal) of mixed fuels covered the surface of the small boat harbor with a layer up to 20 cm thick (Figure 4). The volume of petroleum products washed up on the south coast was not estimated. Only small quantities of oil reached the shore on the Wilkes Island coast to the northwest of the harbor entrance and relatively few dead fish were seen stranded on that shore.

Large numbers of dead fish were stranded mostly along two high-water levels between the harbor entrance and Peacock Point. The odor of putrefying fish was strong as far as 3.2 km away. The intensity of the kill diminished

Figure 2.--R. C. Stoner, aground off the harbor entrance.

Figure 3.--Oil blackened surf.

Figure 4.--Inner small boat harbor was covered with about 20 cm of oil.

along the shore towards Peacock Point. On the southern side of the point, few dead fish were seen, and no dead fish or shoreside petroleum pollution was seen on the northern side of the point.

On September 11 and 12, FAA personnel in cooperation with crewmen from the R. C. Stoner had cleared most of the larger fish from the shore area that had received the bulk of the dead fish. The cleared shoreline extended from the harbor entrance southeast for about 2,300 m (Figure 1).

By September 13, a U.S. Navy harbor clearance team had arrived from Subic Bay, Philippine Islands, to assist in removing the vessel. Two Navy tugs, a Navy salvage ship, the U.S. Conserver, a Navy tanker, the U.S. Noxubee, and USCG Mallow were laying offshore to assist in the salvage of the R. C. Stoner and her cargo.

Standard Oil Company and U.S. Navy personnel were removing oil products from the surface of the small boat harbor. Utilizing air-driven pumps and surface skimmers, they were pumping oil into pits dug close to the harbor. The oil in the pits was burned each evening. Over 100,000 gal were estimated to have been removed from the harbor and disposed of by this technique.

SHORELINE SURVEYS

We made spot surveys along the full length of the seaward and lagoon coastlines of the atoll to determine and assess the effects of the contamination.

The Lagoon

No petroleum products had entered the lagoon. The harbor is blocked from the lagoon by an earthen causeway which had prevented entry of oil into the lagoon from the harbor. The prevailing westerly winds and currents evidently prevented pollutants from entering the lagoon over the reef on the northwestern end of the atoll.

Area Cleared of Fish by the Federal Aviation Agency

With few exceptions all the larger fish had been removed from the cleaned section of the shoreline. The remaining small fish were concentrated in windrows along two oily high-water marks along the coastline. Both high-water lines were the result of abnormally high tides in combination with the strong onshore wind and high seas which prevailed early in September. They were about 8 and 3 m higher up the beach than the high-water level on September 13. The fish were usually thickly covered with oil. They consisted largely of pomacentrids (damselfishes) and acanthurids (surgeonfishes); Pomacentrus nigricans, Abudefduf sordidus, and Acanthurus triostegus seemed to predominate. We estimated that the remaining fish on this section of the coast probably did not amount to more than a few hundred kilograms.

Dead turbine molluscs, Turbo sp. (Figure 5), and dead sea urchins, Tripneustes sp., were abundant. We also saw a few dead beach crabs and small cowries.

Personnel involved in the fish cleanup oper-

4

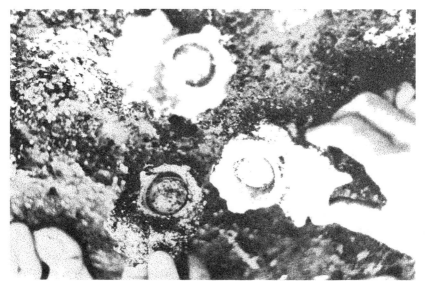

Figure 5.--_Turbo_ sp. suffered high mortality.

Figure 6.--Fishes stranded on the oil blackened beach.

ation had saved specimens of the various species they had collected. These were identified and photographed. Their records showed that approximately 1,360 kg of fish were collected during the cleanup. Based on their rough estimates, by number, 40% were surgeonfishes, mostly Acanthurus triostegus and A. achilles; 30% were parrotfishes; 10% were squirrelfish Holocentrus lacteoguttatus; and 20% were other species, many of which were groupers. Very few dead moray eels were collected during the cleanup.

They saw numerous dead sea urchins and turbine shells, but these had not been removed. A visit to the dump, where the fish had been disposed of, enabled us to verify some of the information supplied, relative to quantities and percentages.

Area Not Cleared by the Federal Aviation Agency

Oil contamination continued for about 2.4 km southeast of the section that had been cleaned. Here too, the majority of dead fish were well above normal high water. Some were even enmeshed in the branches of low bushes. We counted the larger fish (over approximately 15 cm), identified them, and took photographs (Figure 6). Table 1 lists the species we identified in the kill, including specimens saved by the cleanup crew. We did not attempt to identify the hundreds of small oil-covered specimens which no doubt included many additional species. Table 2 includes only the larger fish that were counted on the uncleared shore. The total weight of fish stranded on this section was probably not much over 900 kg.

The scarcity of moray eels was interesting. Local divers and fishermen said that morays were abundant on the reef, but only two dead moray eels were found. During the underwater surveys of the reef flat none were seen. The eels may have detected the pollution and fled the area before the concentration became lethal, or they may have survived it and were not seen.

We saw considerable numbers of dead turbine shells. No counts were made. Few sea urchins had been killed along this section compared with the more northwesterly shore. The only other dead invertebrates were occasional cowries, nudibranchs, and grapsoid crabs. Numerous hermit crabs were observed feeding on the dead fish, but we saw no dead hermit crabs.

The Small Boat Harbor

The small boat harbor was a trap for large quantities of the petroleum spillage. There was a layer of mixed fuels which at times was 20 cm deep in the inner part of the basin. The harbor banks were fouled with a thick layer of

Table 1.--Some of the fishes washed up on the south coast of Wake Island, east of the wreck of R. C. Stoner.

Serranidae (groupers)
    Cephalopholis argus
    At least two other unidentified species

Holocentridae (squirrelfish)
    Holocentrus lacteoguttatus

Scaridae (parrotfishes)
    Scarus perspicillatus
    Scarus sordidus
    At least two other unidentified species

Mullidae (goatfish)
    Unidentified

Acanthuridae (surgeonfishes)
    Acanthurus achilles
    Acanthurus nigricans
    Acanthurus triostegus
    Ctenochaetus striatus
    Zebrasoma flavescens
    Zebrasoma veliferum

Chaetodontidae (butterflyfishes)
    Centropyge flammeus
    Chaetodon lunula
    Chaetodon semeion

Balistidae (triggerfishes)
    Melichthys vidua
    Rhinecanthus rectangulus

Scorpaenidae (scorpionfish)
    Pterois volitans

Diodontidae (puffer)
    Diodon hystrix

Muraenidae (moray)
    Unidentified

Mugilidae (mullets)
    Unidentified

Pomacentridae (damselfishes)
    Abudefduf imparipennis
    Abudefduf sordidus
    Pomacentrus nigricans

Priacanthidae (bigeye scad)
    One specimen found--unidentified

Cirrhitidae (hawkfish)
    Unidentified

Carangidae (jack)
    Unidentified

Labridae (wrasse)
    Thalassoma umbrostigma

Table 2.—Fishes (>ca. 15 cm) counted on the shoreline which had not been cleared by the FAA.

Family	Count	Percentage of total number	Percentage of total estimated weight
Holocentridae (squirrelfishes)	384	37	9.5
Acanthuridae (surgeonfishes)	288	27.4	7.6
Scaridae (parrotfishes)	284	27.4	59.9
Serranidae (groupers)	52	5.0	21.0
Balistidae (triggerfishes)	20	2.0	1.0
Scorpaenidae (scorpionfishes)	8	0.6	0.2
Mullidae (goatfishes)	4	0.3	0.1
Diodontidae (spiny puffers)	4	0.3	0.2
Muraenidae (morays)	2	0.1	0.1

black bunker fuel. A survey of the kill along the perimeters of the channel and inner harbor revealed considerable numbers of small fish, very few large fish, numerous grapsoid crabs, and some small sea snakes; most of these were heavily fouled with black oil. The fish included various pomacentrids, small soarids (parrotfishes), and acanthurids. Although the fish were numerous, the total biomass of the kill was relatively small, probably only a few hundred kilograms. We learned from FAA personnel who worked around the docks that the fish population in the harbor was normally sparse. The heavy layer of oil on the surface and emulsified jet fuel below precluded diving in the channel or harbor; however, survival of marine vertebrates and invertebrates seemed unlikely under such intense contamination.

Sea Birds

Hundreds of thousands of sooty terns were aggregated in the air to the northwest of Wilkes and Peale Islands (Figure 7); frigate birds in considerable numbers and a relatively few shearwaters were also present. The birds were concentrated over areas which did not

Figure 7.—Large aggregations of sooty terns off the western end of the island.

receive petroleum spillage. No birds were seen along the fouled coastline. There was no indication that any bird life on the atoll had been killed or harmed by the spill.

UNDERWATER SURVEYS

Underwater surveys were made (1) in the vicinity of the R. C. Stoner, (2) 2.5 km southeast of the harbor entrance, (3) on the reef flats and reef fronts about 1 km southeast of the harbor entrance, and (4) on Wilkes Island about 300 to 400 m northwest of the R. C. Stoner (Figure 1). Dives were also made in the lagoon and in uncontaminated areas along the northwesterly reefs. The lagoon supports a large population of fish. Had large quantities of petroleum products entered this shallow enclosed area mortality would probably have been very high. No dead fish were seen in the water at any of these places.

The Vicinity of the Wreck

During survey 1 the ocean around the R. C. Stoner contained a considerable amount of what was probably jet fuel that had emulsified with the water which was a dirty straw color to a depth of 7 m or more. This contamination was sufficient to cause skin irritation and after an hour we were covered with a light oily film and were in much discomfort. Itching and tenderness of the skin in the more sensitive areas persisted for several days.

Many fish were seen on the reef around the wreck. Most of them were on the bottom (10-12 m) where the water was less contaminated. Many were within 2-3 m of the ship. Curiously, we saw numerous individuals of the leatherback runner, Chorinemus sanctipetri (Figure 8), frequently come up and swim around in the upper 3 m of water where the pollution was heaviest. At the after end of the vessel the water was so contaminated that we dared not swim around the stern. In spite of this, we saw leatherback runners swimming in this area, with no apparent ill effects. We thought that the combination of irritation and oiliness would have a deleterious effect, particularly on the gills. However, individual fish may have remained in the contaminated water for brief periods only.

The coral heads immediately seaward of the wreck and from 3 to 15 m from it were well populated with fish. These included parrot-

Figure 8.--Leatherback runners, Chorinemus sanctipetri, swimming in an area heavily contaminated with jet fuel.

8

fishes (Scarus perspicillatus, Chlorurus gibbus, and at least two other species, probably S. sordidus and S. brunneus), surgeonfishes (Acanthurus triostegus, A. achilles, A. nigricans, and A. guttatus), butterflyfishes (Chaetodon setifer, C. auriga, C. lunula, C. quadrimaculatus, and Forcipiger longirostris), grouper (probably Cephalopholis argus), porgy (Monotaxis graduculis), wrasse (Thallassoma umbrostigma), triggerfishes (Melichthys vidua and M. buniva), puffers (probably Diodon hystrix and Arothron meleagris), damselfishes (Pomacentrus nigricans and Dascyllus aruanus), and squirrelfish (probably Holocentrus lacteoguttatus). Two unidentified species of jacks (Carangidae) were numerous.

Although there were numerous fish around the reefs close to the ship, we later found, after subsequent dives, that other areas on the reef front farther away from the source of pollution were more densely populated than the reef front near the R. C. Stoner. During the period of greatest petroleum spillage, many fish in the wreck area were probably either killed or driven away.

The reef flat directly inshore of the wrecked ship is about 0.5 to 2 m deep at low tide and about 125 m wide, the widest such area on this coast. We found it nearly barren of fish. The bottom is flat, covered with coralline rubble with few coral heads. One would not expect such a reef normally to support a very large population of fishes, but neither would it be expected to be as depauperate as it was. Probably, many of the fish in this area had been either killed or driven off.

Sea Off Wake Island Between the Harbor and Peacock Point

Time and weather permitted us to make two underwater surveys off this coast (surveys 2 and 3, Figure 1). Both surveys were made in areas where considerable numbers of fish had been stranded on the shore.

On September 15, the wind was still strong from the southwest. A heavy sea was breaking on the reef 2.5 km southeast of the harbor, where we made survey 2. Weather reports indicated that typhoon Sarah would pass very close to Wake in about 48 hr and sea conditions were poor for diving. However, we decided it would be best to survey the polluted coast outside the reef as best we could before the typhoon arrived. We were unable to find any passes through the reef so scuba was not used.

The heavy turbulence on the reef flat made visibility poor. The reef flat there is about 30 m wide and has numerous coral formations. It should support a substantial fish population; however, we saw relatively few fish. The dominant groups were various scarids (not including adult humpback parrotfish, C. gibbus), and acanthurids, the latter mostly A. triostegus and A. achilles. We saw no grouper or squirrelfish. There was no petroleum or dead fish in the water. At least part of the sparsity of fish on the shelf was probably due to the heavy seas.

The reef front drops off steeply to a depth of about 8 m. Visibility was better than on the flat. There were numerous rock and coral formations and a myriad of fish. Time did not allow us to obtain any data on the density of the fish population on the seaward side of the reef. However, on the basis of past experience, we considered the population to be high. This population included a wide range of the common reef groups. The most numerous species were generally the same ones that were the most plentiful in the kill: several species of parrotfishes; surgeonfishes, predominantly A. triostegus and A. achilles; several species of groupers, mostly C. argus, and various pomacentrids and chaetodontids. Exceptions were squirrelfish, few of which were seen, and adult humphead parrotfish, which was numerous outside the reef but apparently absent in the kill.

Shortly after crossing the reef, we encountered numerous blacktip sharks, Carcharhinus melanopterus. During this survey and subsequent surveys outside the reef, we were frequently pestered by sharks, most of them 1.2 to 1.5 m long. They were extremely curious and persisted in making close passes, frequently coming within less than 1 m of us. They were remarkably bold and did not scare easily. Local skindivers considered the abundance and boldness of the blacktips as very unusual. The only way we can account for their concentration and behavior is that the sharks had been attracted to the area by dead fish and had become conditioned to finding abundant food drifting around. Such conditioning might result in immediate attack behavior, without the preliminary investigation usually more characteristic of sharks. The behavior of the sharks was a little disconcerting and hindered us somewhat, as we had to keep a constant watch in order to chase them off. It is interesting to note that no shark was reported in the kill.

9

We covered about 275 m of coastline. As on the reef flat we saw no dead fish or invertebrates, nor any indication of petroleum in the water.

We made survey 3 off the south coast about 1.7 km southeast of the wreck (Figure 1) on September 16. The wind had abated and the sea was the calmest we had seen since our arrival. The heaviest beach-stranded kill had occurred in this general area. We surveyed the reef flat for about 125 m. The reef in this area was similar to that in survey area 2, but the fish population, dominated by Scarus sordidus, Acanthurus triostegus, A. achilles, and Ctenochaetus striatus, was larger. We saw no squirrelfish, grouper, or adult humphead parrotfish on the reef flat. The reef front dropped off abruptly to about 8 m. We were again impressed by the abundance of fish life, which was far greater than on the reef flat. Here also, parrotfishes and surgeonfishes dominated. We saw several different species of Scarus and numerous large C. gibbus; in addition, C. argus and at least three other species of groupers and two species of carangids were plentiful. We saw a number of squirrelfish, which appeared to be the same species, H. lacteoguttatus, which occurred in the kill. As in survey 2, we saw most of the species that were in the kill, plus many more. Those species which were most abundant in the kill were apparently also the most abundant offshore. Humphead parrotfish and squirrelfish were exceptions. We encountered numerous blacktip sharks behaving in the same manner as on the previous day.

Sea Off Southeastern End of Wilkes Island

The area in front of the Pacific Missile Range facility about 300 to 400 m northwest of the R. C. Stoner was chosen for survey 4 (Figure 1). Because of the prevailing southwesterly winds and possibly the current, very little oil had washed up there, but instead had been driven up the channel into the inner harbor. On our survey of the beach on Wilkes Island, we had found little oil and few dead fish. Because that area had apparently been relatively little affected, we thought it would serve as a useful comparison to the two more easterly surveys. Local scuba divers told us it was one of their favorite diving spots, and that there were many fish on the reef front and sharks were rare. Curiously, during our brief survey, we found neither the reef flat nor the reef front nearly

as abundant in fish as the previously investigated areas; we saw few parrotfishes and no dead fish in the water, but we were bothered by blacktip sharks and later by two larger (2.0-2.5 m) gray sharks. The latter, which we could not positively identify as to species, manifested the same aggressive tendencies as the blacktips, making very close passes. We were a bit more leery of large sharks with this type of behavior and left the area posthaste.

CONCLUSIONS

By noon of September 16, large storm seas were rapidly building up on the reefs and it was impossible to continue the survey. That night typhoon Sarah struck Wake with winds up to 67 m per sec (140 miles per hr) causing great damage to the island's facilities. The typhoon, however, had one saving grace: it blew away virtually all of the oil that had accumulated in the inner harbor and even did a good job of scouring its oil-fouled banks, solving in one night a problem which would have taken many weeks to overcome. On the morning of September 17, the harbor was clear and clean. A brief survey of the affected beach areas disclosed that the only remaining evidence of pollution was black oil embedded in reef flat crevices and impregnated in coral. The authorities on the island were evacuating all nonessential personnel because of an acute housing shortage. Thus, we were unable to make an underwater survey of the now oil-free inner harbor. This was unfortunate. Even though we have no "before pollution" data on the harbor, it would have been of great value to have an accurate assessment of the biological conditions of a shallow confined area of this nature, the surface of which was completely covered with a thick film of heavy oil and the subsurface contaminated with light fuel for over a week.

We estimate that about 2,500 kg of dead fish were washed up on the south shore during the period of maximum petroleum escapement, the first week after the R. C. Stoner had foundered. Our best guess is that most of the kill had occurred on the shallow reef flat bordering the coast. The available evidence appears to support this theory. Numerous groupers and squirrelfish occurred in the kill. Apparently, all of the squirrelfish were H. lacteoguttatus and nearly all the groupers were C. argus. We saw neither of these groups on any of the reef shelves we surveyed, yet they are generally the most abundant of their respective families occurring in shallow water around Wake. C. argus were frequently observed outside the reef. We saw few H. lacteoguttatus outside the reef; however, as with the eels, assessment of a squirrelfish population in the daytime by visual means is impossible, especially without using scuba. Groupers and squirrelfish on the reef shelf may have holed up when they sensed the pollution, rather than escaping to uncontaminated water outside the reef. Thus, they would be very vulnerable and suffer exceptionally high mortality. As far as we could determine, no adult humphead parrotfish, C. gibbus, was killed. Adults of this species were not seen on the reef shelves that had been polluted nor in unpolluted shallow areas but were numerous outside the reef. It is possible that young C. gibbus were among the kill and not detected. C. gibbus does not develop the enlarged forehead characteristic of the adult until it is quite large. During the survey we were not familiar with the appearance of the young fish. Other species of parrotfish (these may have included young C. gibbus) were apparently very vulnerable to the petroleum pollutants. If one assumes that adult C. gibbus were also vulnerable, lethal pollution probably did not extend into the deeper water outside the reef.

It is interesting to speculate as to which of the various fuels caused the most mortality. Within an hour after the grounding it was known that aviation gasoline was escaping because its distinctive smell permeated the area and eventually spread over most of the island to leeward of the vessel. However, concurrently with the gasoline, J-P4 jet fuel and possibly A-1 turbine fuel were leaking. There is no record of the extent of fish mortality or even if any mortality had occurred the day the ship grounded. The morning after the grounding (September 7), black fuel oil was leaking in large amounts and gross black oil pollution extended along the shoreline and into the small boat harbor. Dead and dying fish were first noted on September 7. On September 8 there was an extensive fish kill along the beach. All of the aviation gasoline had apparently spilled by the morning of September 9 as the presence of gasoline vapor in the air could no longer be detected. A large part of the black oil had also spilled by September 9, but the remainder continued to escape at least through September 15. Large quantities of aviation jet fuel, aviation

turbine fuel, and possibly diesel fuel were escaping through September 16. However, there was no evidence of fish mortality subsequent to September 10.

Near the wreck we observed that numerous reef fishes and leatherback runners showed a surprising tolerance to jet fuel or turbine fuel, or both. The seaward side of the reef, below the surface, was probably contaminated only with light fuels, including aviation gasoline, whereas the reef flat received both light fuels and black oil mixed into the water by the heavy surf.

Although admittedly poorly documented, the evidence indicates that the petroleum product most lethal to fish was either aviation gasoline (which contains tetraethyl lead, a known toxic agent) or black oil. Other oil spillages have indicated that fish are not seriously affected by crude oil, which is primarily confined to the surface. Laboratory experiments have corroborated this finding. However, when a large volume of heavy oil is well churned onto a shallow reef flat, the mortality may be expected to be high.

Considerable numbers of fish must have been killed on the reef flat and washed out into deep water; however, we were unable to estimate the percentage of the kill that did not wash ashore. By September 13, when we started our investigation, fish apparently were not being affected by oil pollution. We saw neither dead fish in the water nor any fish showing signs of distress.

We made only cursory observations on the invertebrates. The turbinid snails and sea urchins were the only animals found dead in large numbers. The population of Turbo on the south coast of Wake may have been seriously depleted. Damage to many of the various small invertebrates commonly inhabiting coral reefs may have been severe. These invertebrates have the slowest recovery rate and thus their depletion would have the most profound effect on the ecosystem of the reef flat.

We observed that the fish population on the reef front and slope was dense. This proximal population would provide a source for rapid recruitment to both the reef flat and the inner harbor.

That the fish kill was negligible when compared with the surviving population may be attributed to three main factors. First, the nature of the coastal terrain was a saving factor, composed as it is of a narrow reef flat with the water depth dropping off rapidly along the reef front. Places with great expanses of shallow tidal area are far more vulnerable to any type of contamination. Secondly, the fact that oil was unable to pass from the inner harbor to the lagoon. Wake Island, of course, does have a relatively shallow area in the lagoon vulnerable to contamination. Had the heavy concentration of fuels which accumulated in the harbor entered the lagoon, a near "total kill" with long-lasting effects probably would have occurred. Third, a large percentage of the spillage entered and was trapped in the small boat harbor where it could do relatively little damage to marine life. Had this oil been able to spread along the outside reefs, the kill would have been greater.

After the typhoon a Navy salvage team found that the wreck had broken into three sections. All her tank tops were open to the sea and apparently all of her cargo had been removed by sea action.

The portion of the wreck remaining above water was considered hazardous to the instrument landing system on the runways, so the wreck was flattened to the waterline with ex-plosives. During the course of the demolition operation there was minor weeping of petroleum products that formed a slick extending about 1.5 km offshore. With northeasterly trade winds prevailing there was no more shoreline pollution.

Poor weather and shortage of the time limited the scope of this survey, but the major shortcomings of the investigation were that we arrived too late to get the full picture of the damage, and that we had no clear concept of the normal pre-pollution situation in the areas we surveyed. We suggest that if at least two trained teams of biologists, including specialists in fish and invertebrates, were located at suitable laboratories on the Pacific and Atlantic Oceans, we could enhance our knowledge of the effects of oil pollution more efficiently. The teams would have suitable survey equipment readily available and could be dispatched immediately to any area where a petroleum pollution threat is anticipated or already exists. If possible, the group would reach the scene in time to survey the environment and biota before the oil arrives.

12

Oil Pollution on Wake Island from the Tanker *R. C. Stoner*, by Reginald M. Gooding, Special Scientific Report—Fisheries No. 636

ERRATUM

On page 9, right column, first paragraph, line 7:
    "humpback parrotfish" should read "humphead parrotfish."

the reef fro
shallow tidal
any type of
that oil was
bor to the
does have a
goon vulner
heavy concer
in the harbo
kill'' with lc
have occurre
the spillage
small boat h
little damage
able to sprea
would have be

After the t
that the wrec
All her tank
apparently al
by sea action.

The portio
water was co
ment landing
wreck was f

CPSIA information can be obtained
at www.ICGtesting.com
Printed in the USA
BVHW071018141218
535632BV00019B/888/P